When Things Change

BOOK 2

Join Yoyo and AJ(Aunt Jessi) as they explore how one simple ingredient can look, feel, and behave in different ways. Through melting, mixing, cooling, and wondering, children discover that change does not mean something disappears — it simply becomes something new.

Science inside

States of Matter • Conservation of Mass

Thinking skills

Observation • Curiosity • Patience • Experimentation

HEERAL SHARMA

Part of the Wonder Loop™ Series
Learning through living, wondering, and doing together.

Published by Happy Orange
myHappyOrange.com
PO Box 266, Tahoe Vista, California

ISBN: 978-1-972521-01-4 (Paperback)

Written by **Heeral Sharma**
Book design by **Purple Parlo**

Illustrations created using **Scrively**
under the direction of the author.

Foreword

Some children learn best by listening.

Some children learn best by watching.

And some children learn best by touching, mixing, smelling, wondering, and discovering.

The Wonder Loop™ Series is created for children who need to experience learning — not just hear about it.

In these stories, science unfolds slowly. There is no rush to find answers. No long explanations. Just shared moments, small questions, and the joy of noticing.

Curiosity grows best when it feels safe.
And sometimes, the most powerful learning begins with

"Hmm... I wonder."

A Wonder Loop™ Book

Read slowly. Pause. Notice. Wonder together.

LOOK ▸ WONDER ▸ ACT ▸ REFLECT

Wonder loop™

AJ unwrapped a bar of chocolate.

Yoyo’s tummy grumbled.

Chocolate meant tasting.

Crunch!

Hard. Smooth. Delicious.

Solid.

AJ placed it over a pot of hot steaming water.

Steam curled up softly.

Poof!!

Slowly... the edges softened.

One slumped.

Then another.

Soon...

Smooth. Shiny. Swirly. Yum!

Liquid.

“Time to make it interesting!” said AJ.

Green pistachios. Red cranberries.

Plip! Plop!

Yoyo stirred.

He waited...

Nothing happened.

Different!

Cranberries stayed... pistachios too.

AJ poured.

Plop! Thick. Heavy.

It spread slowly.

"Ow!"

Hot... Sticky... Sweet.

Liquid!

Tick! Tick! Tick!

Steam faded. Chocolate cooled.

Solid.

Hard. Crunchy. Chewy.

“Same chocolate — but different.”

The kettle whispered.

Hisss...

Steam floated.

Yoyo tried to catch it.

Nothing.

It just floated.

Gas.

“Tea Time!” said AJ.

Yoyo added honey.

Drip! Mix!

Warm. Sweet. Cozy.

“What if?”

Yoyo smiled.

Mix! Mix! Mix!

"Hot chocolate tea!
My new creation!"
exclaimed Yoyo.

Our world is full of wonder.
Same things.
Different ways of being.
And that's okay.

“Are you ready to give it a try?”

Follow the recipe on the next page.

AJ'S Simple Chocolate Bark

Wonder loop™

What happens when the chocolate warms up or cools down?
Do the toppings behave the same way?
Try adding different toppings and see what happens!

Warning:
Melted chocolate can get very hot and may cause burns. Adult supervision is recommended.

You will need:

- 1 cup chocolate chips or 1 chocolate bar (broken in pieces).
- 2 tablespoons chopped nuts, pretzels, cranberries, or toppings of your choice.
- Parchment paper

Steps:

1. Heat the chocolate for 30 seconds in the microwave, or melt it gently using a double boiler.
2. Stir. Stir.
3. Repeat until the chocolate melts and becomes smooth.
4. Sprinkle nuts, cranberries, pretzels, rice crispies, or sprinkles.
5. Mix. Mix.
6. Pour the melted chocolate onto parchment paper. Use a spoon to spread it into a thin layer.
7. Let it cool. Leave it on the counter until the chocolate hardens again.
8. Break and enjoy. Once it is solid, break the chocolate into pieces.

Touch... Smell... Taste... Share... Enjoy!

More Things to Wonder About

You can extend the experience in your own kitchen by **warming and cooling different things** and noticing what happens.

Try:
A piece of butter in a warm pan.
Then put it back in the refrigerator.

Peanut butter on warm toast.
What happens as it warms?
What happens as it cools?

Syrup poured on warm pancakes.
What happens as it warms?
What happens as it cools?

A bowl of gelatin (Jell-O).
What happens when it warms?
What happens when it cools again?

Ask:
What happens when things warm up?
Do they melt, soften, or stay the same?
What happens when they cool down again?
Can they go back to how they were before?
Do all things change in the same way?

LOOK WONDER ACT REFLECT

wonder loop™

Wonder Moment

Keep wondering.
Your kitchen is full of surprises.

The Science inSide

If your child is not yet asking,
"Why?",
you may want to wait before reading ahead.

Re-read this book together several times.
Lean into the surprise:
"What? No way! What happened?"

When the world feels a little magical,
children begin to wonder why things happen.

Curiosity comes first before understanding.

When children start asking questions, they are ready
for the science behind what they observed.

Solid

A chocolate bar is solid. Inside, tiny parts are packed closely together.
They hold on to each other and stay in place.
That's why solids keep their shape.
You can hold them. Break them. Stack them.

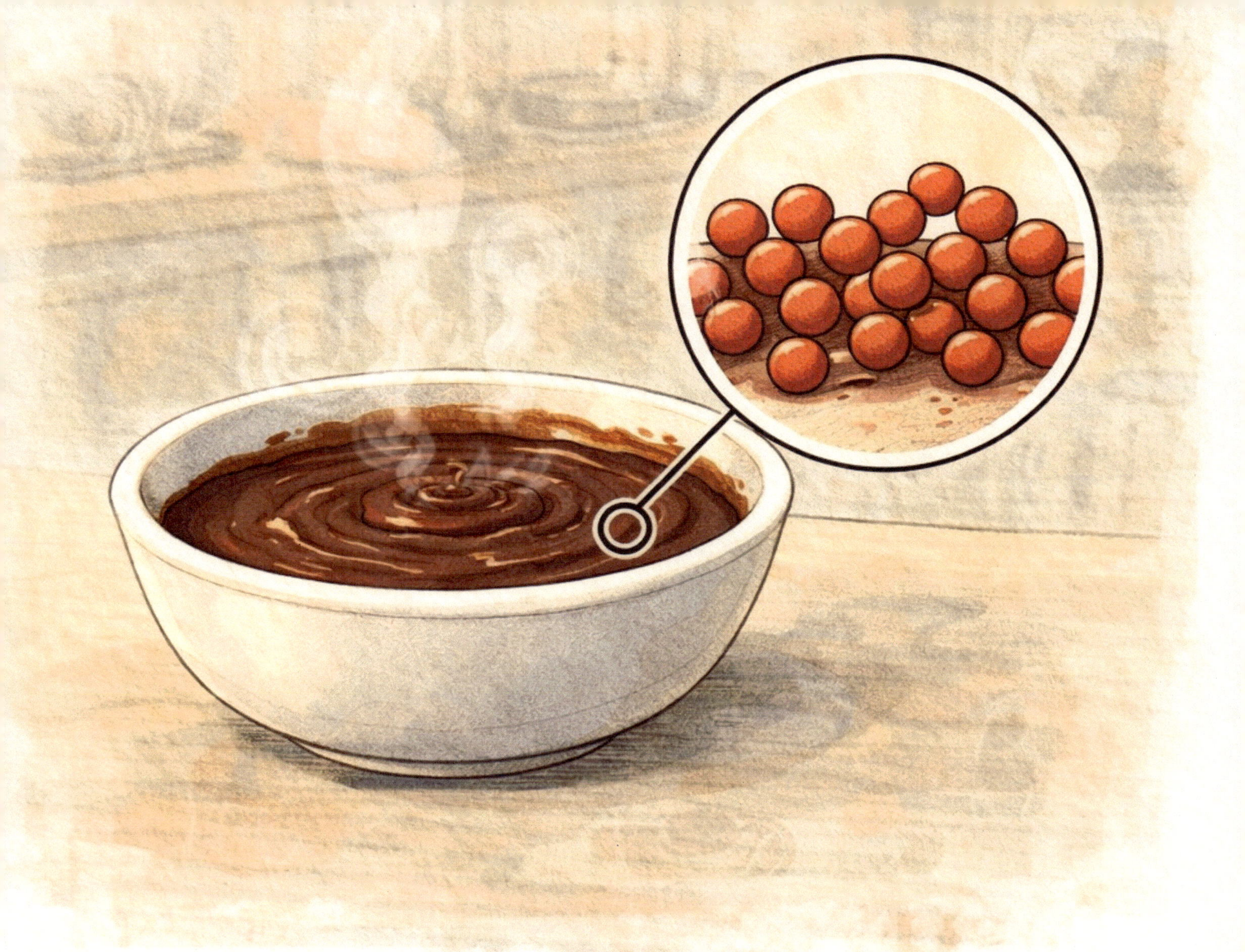

Liquid

When chocolate warms up, its tiny parts begin to move faster.
They are still close together, but now they can slide past each other.
That's why liquids flow.
They can be poured, stirred, and spread.

Gas

When something gets very hot, its tiny parts move very fast and spread far apart.
They are no longer close together.
That's why gases are hard to hold or see.
They move freely and fill the space around them.

Why do different foods change in different ways?

Different materials are made of different parts.
Some parts let go easily when heated. Some parts stay together longer.

That's why, when heated:
chocolate melts, cheese softens, bread toasts and marshmallows puff

Under heat, different things change in different ways.

Big idea!

Things don't disappear when they change.

Their tiny parts just move, spread, or rearrange.
Solid. Liquid. Gas.

What else around you is changing?

A Wonder Loop™ Book

The Wonder Loop™ is a **simple thinking framework** for finding steady ground when the world feels too big or too fast.

1. Pause

Slow down.
Take a deep breath.

2. Look

What do you notice?
What do you see, hear, smell, taste, or feel?

3. Wonder

What do you think is happening?
Ask: Who? Where? When? How? Why?

4. Act

What are my choices?
What can I do next?

5. Reflect

What happened? What did I learn?
What can I try next?

Rooted in RDI® principles of slowing down, using minimal language, and creating space for thinking, it gently guides children to engage in each activity mindfully, building strong episodic memories of competence. It nurtures curiosity, encourages exploration, and supports thoughtful choices. Over time, these small moments of noticing and wondering grow into confident thinking and deeper understanding.

When you are done reading, encourage your child to keep wondering, keep trying, and make a little magic of their own.

Made in the USA
Monee, IL
13 June 2026

54103699R00024